I0480591

Abstract Patterns

GEOMETRIC COLORING BOOK

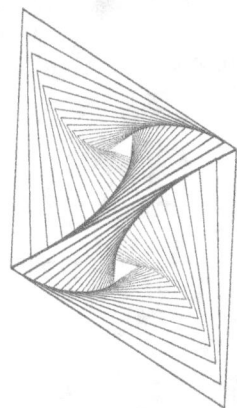

COLORING TIP:

You might find that the spine of the book might get in the way of you coloring. Our suggestion is to cut out each design so that you work on a flat surface. If you are using colored markers or even water color, I would also suggest to have scrap paper beneath your coloring sheets.

We hope you have a peaceful and enjoyable coloring experience!

piggybackpress.com

Copyright © 2021. All Rights Reserved.

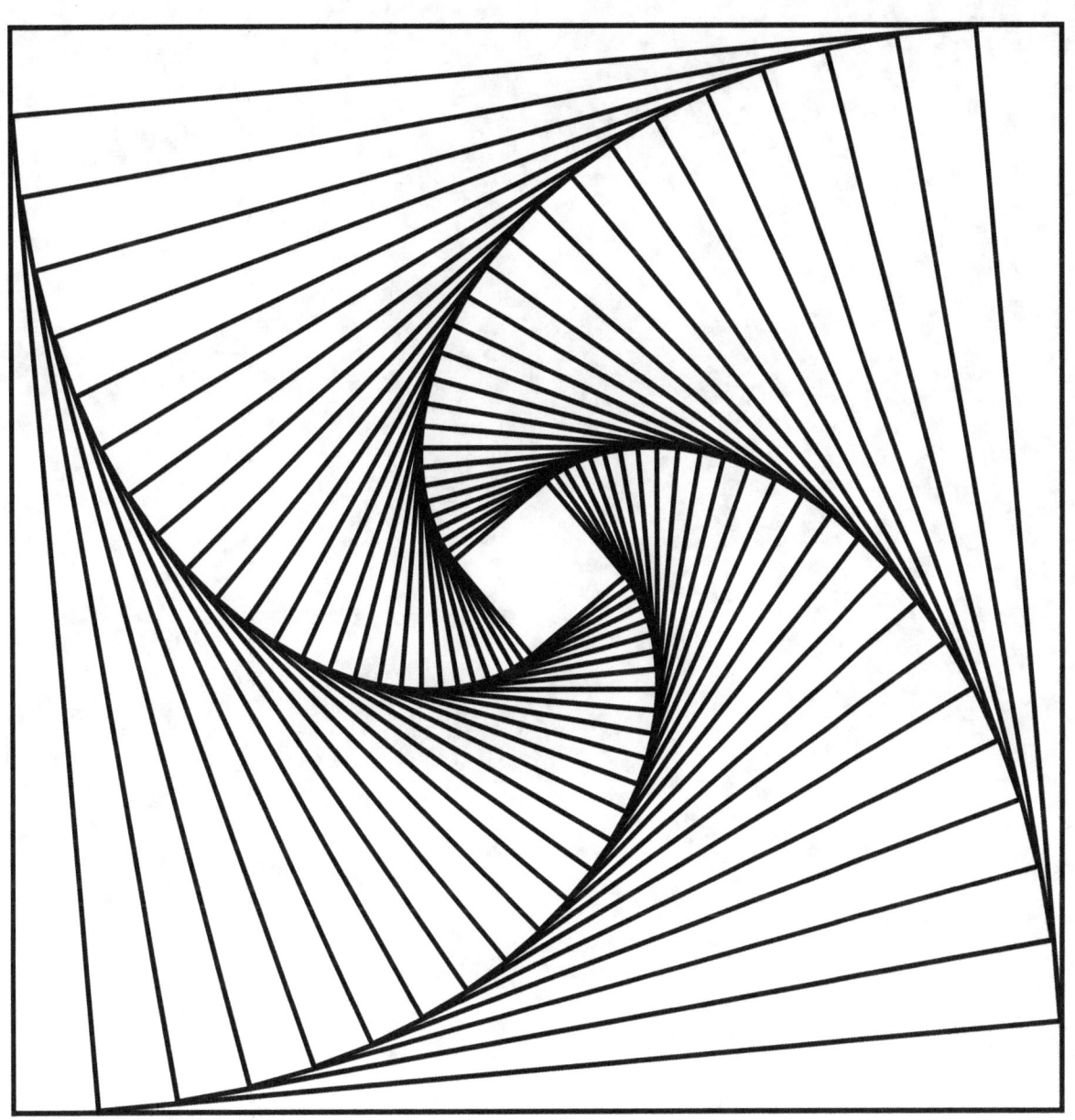

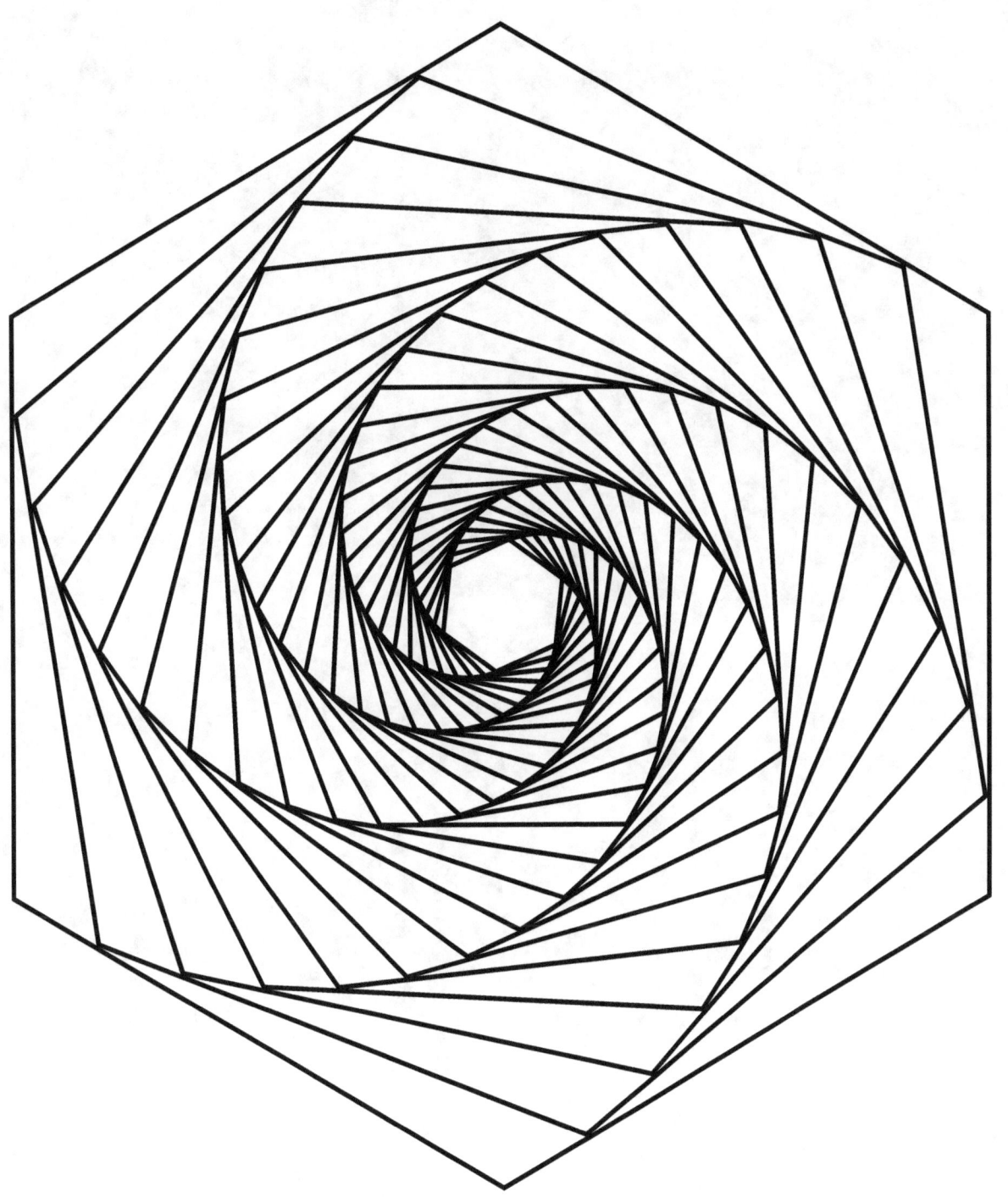

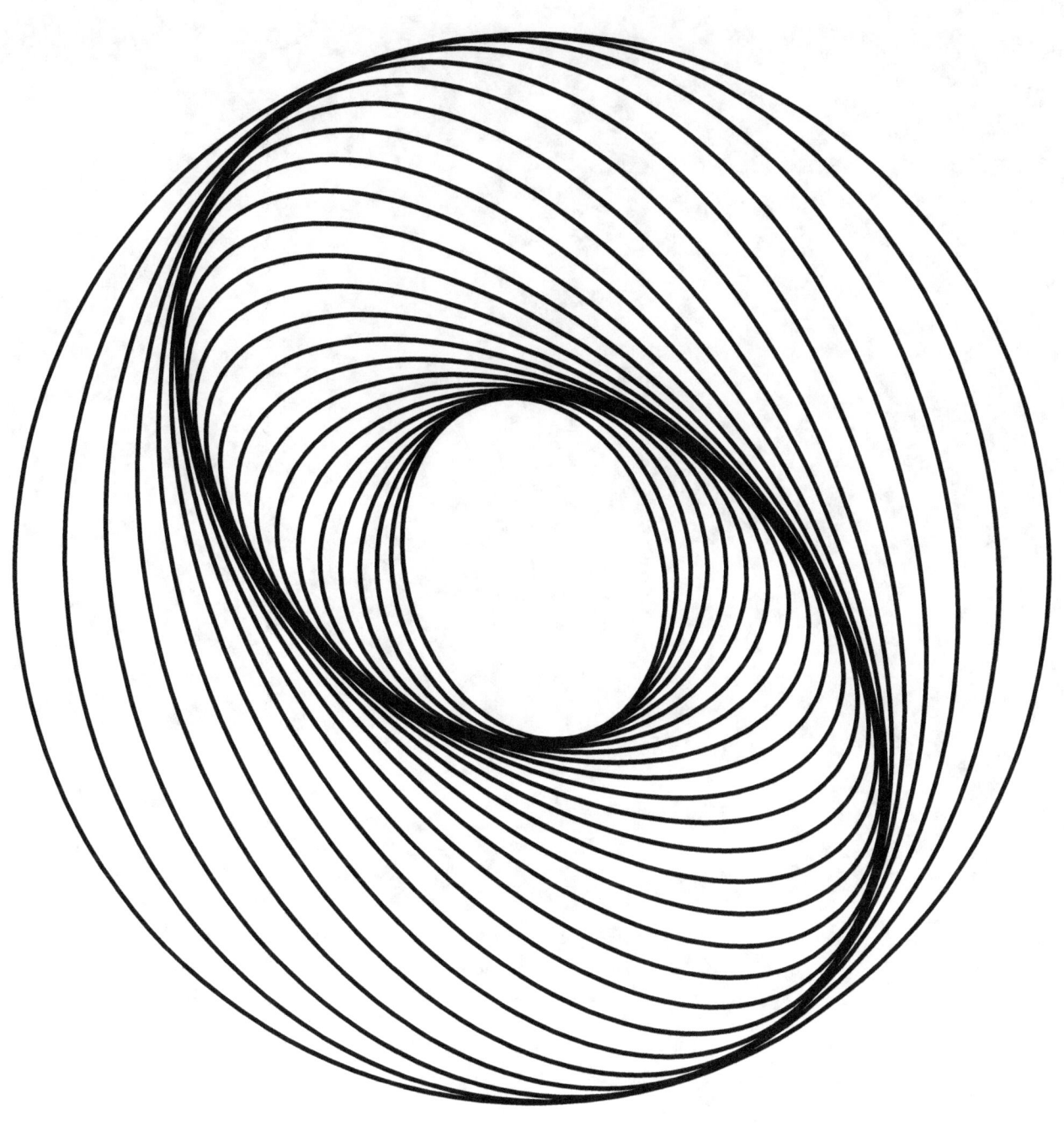

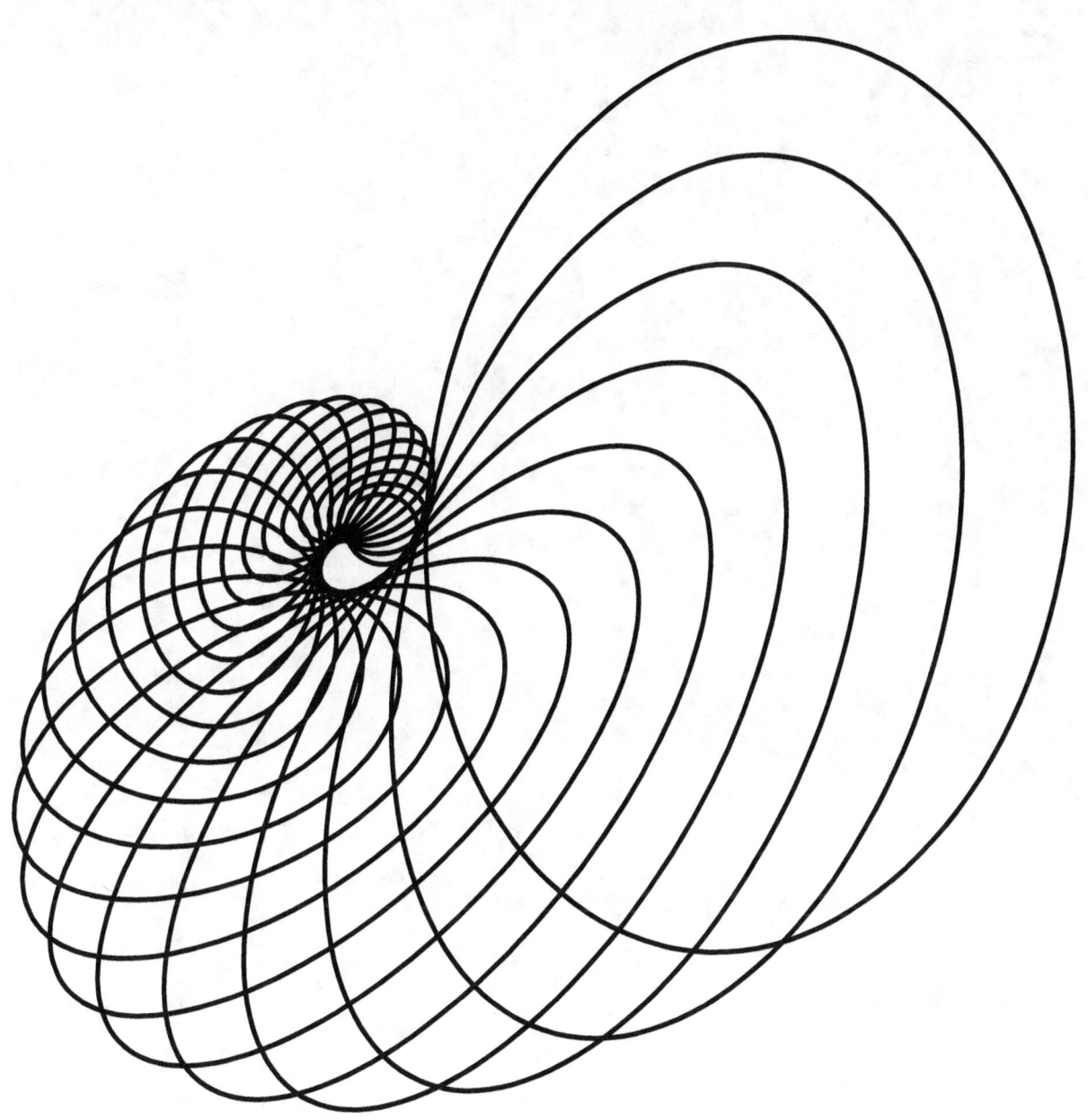

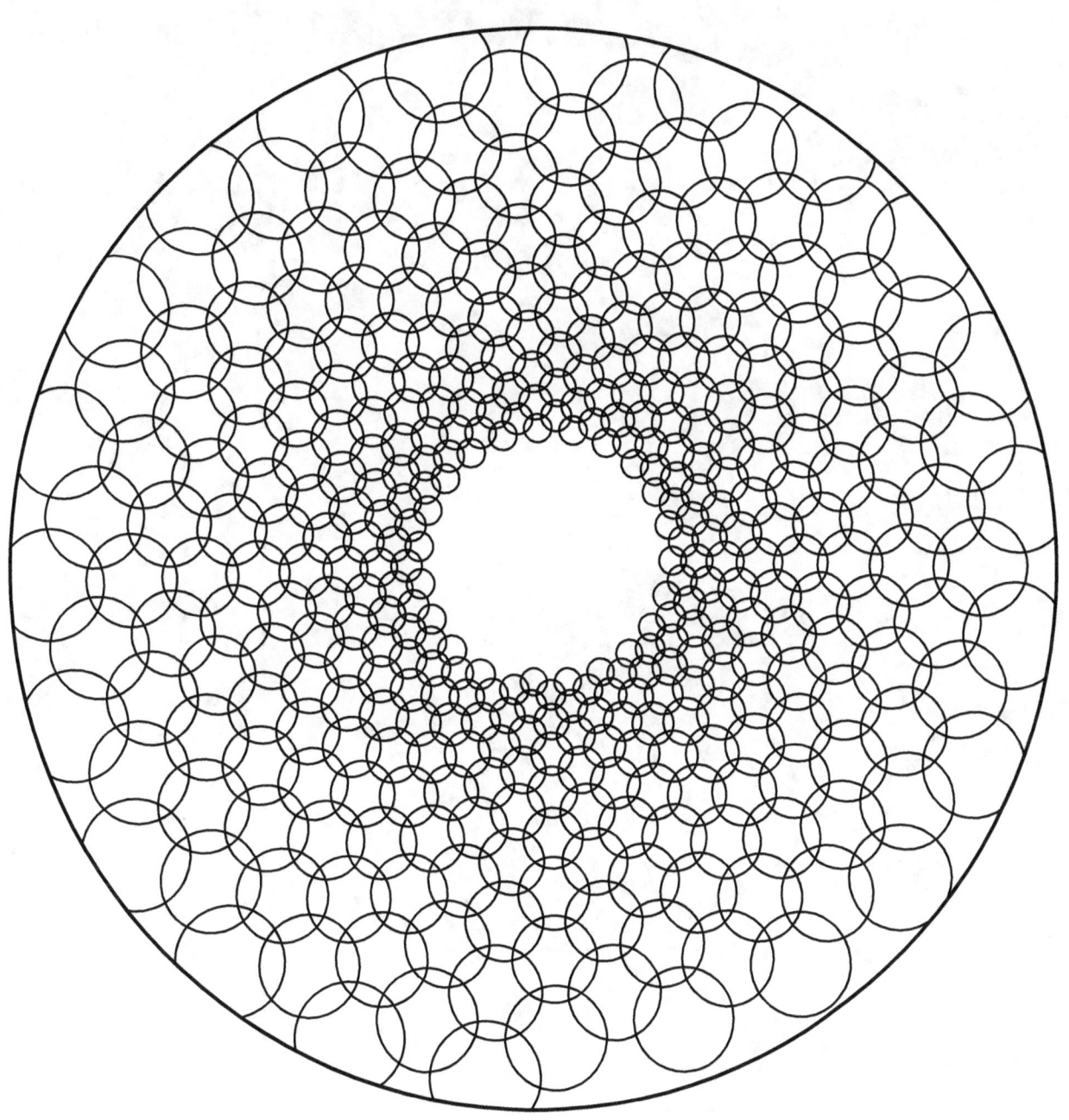

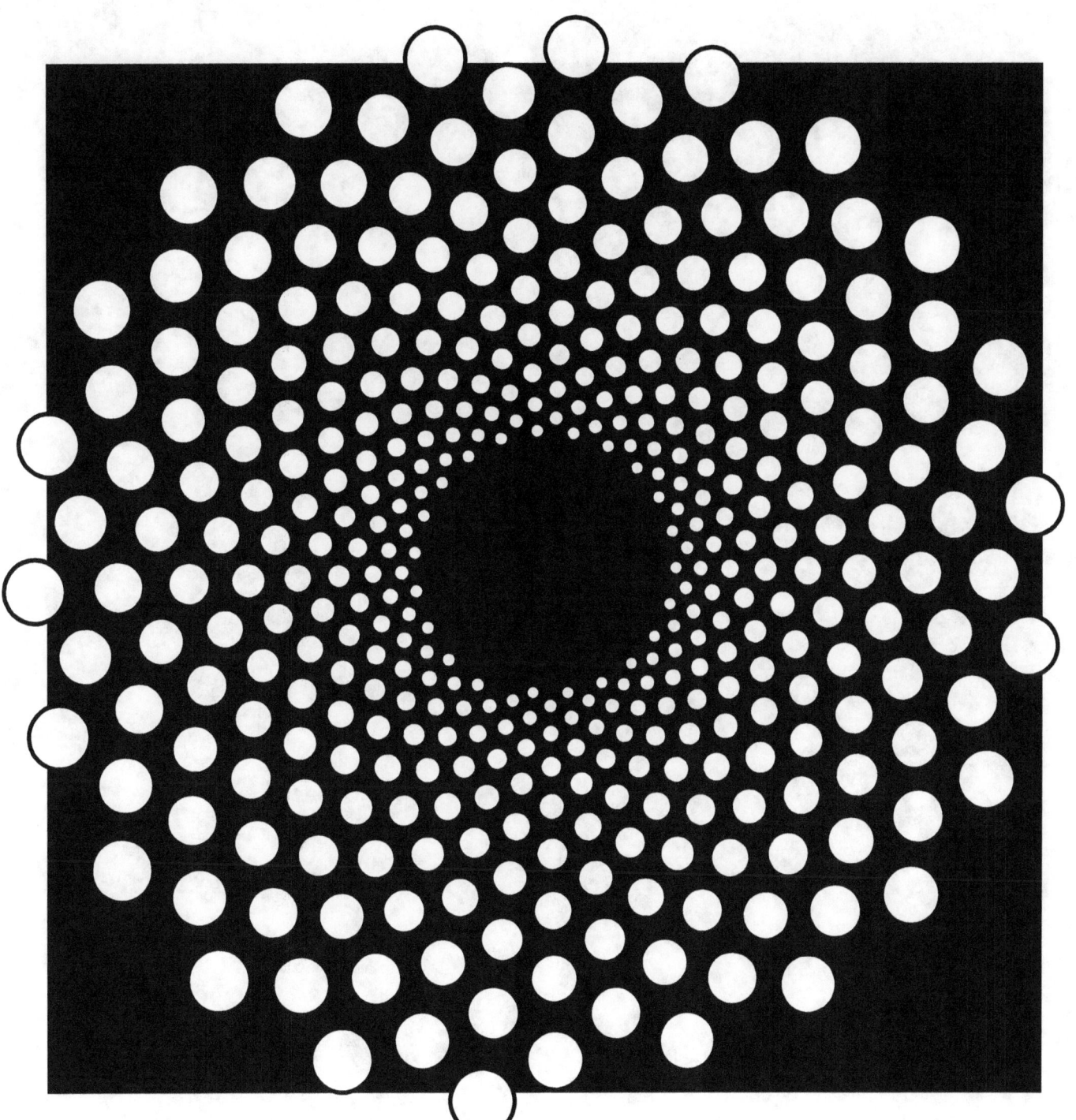

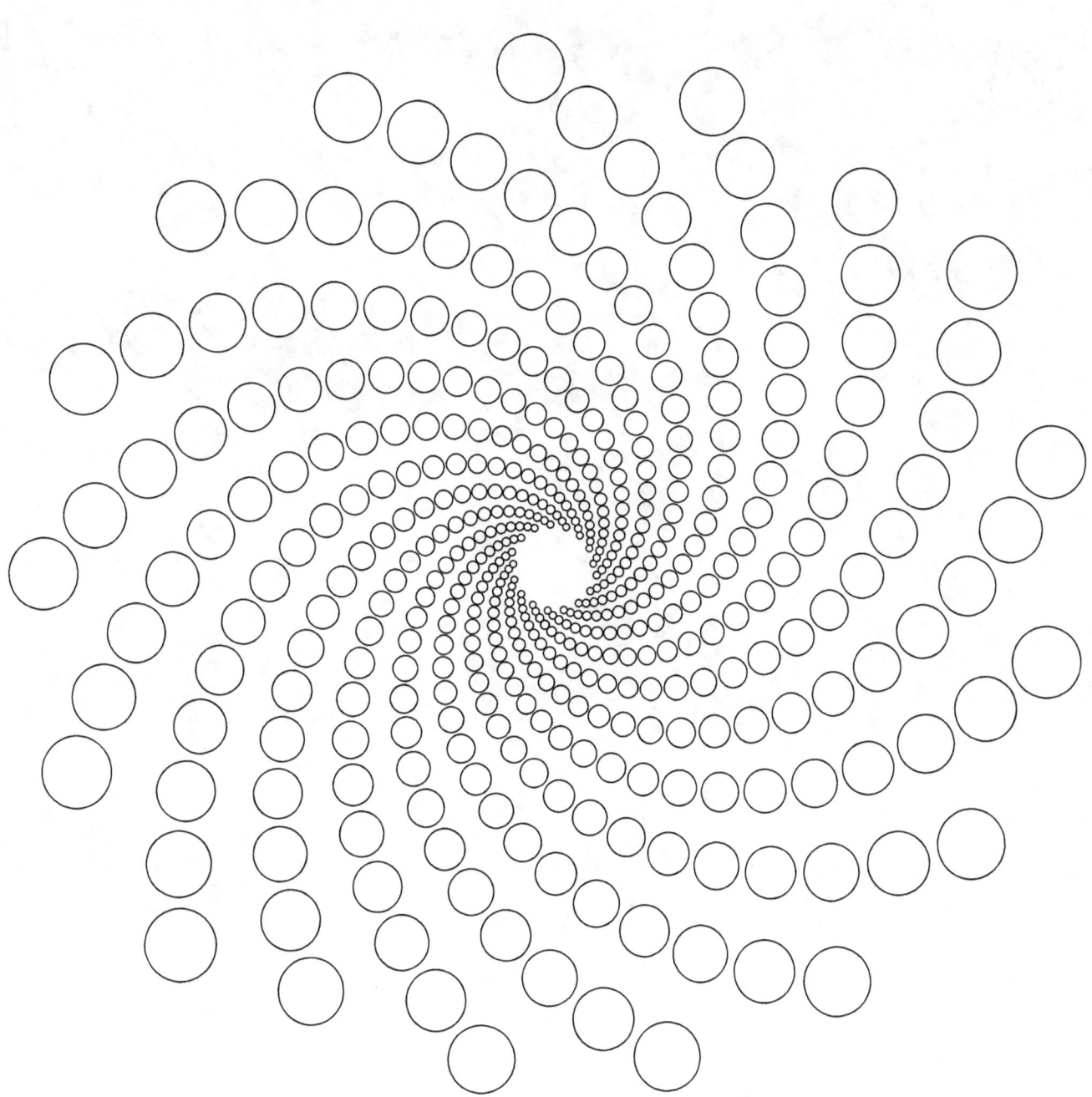

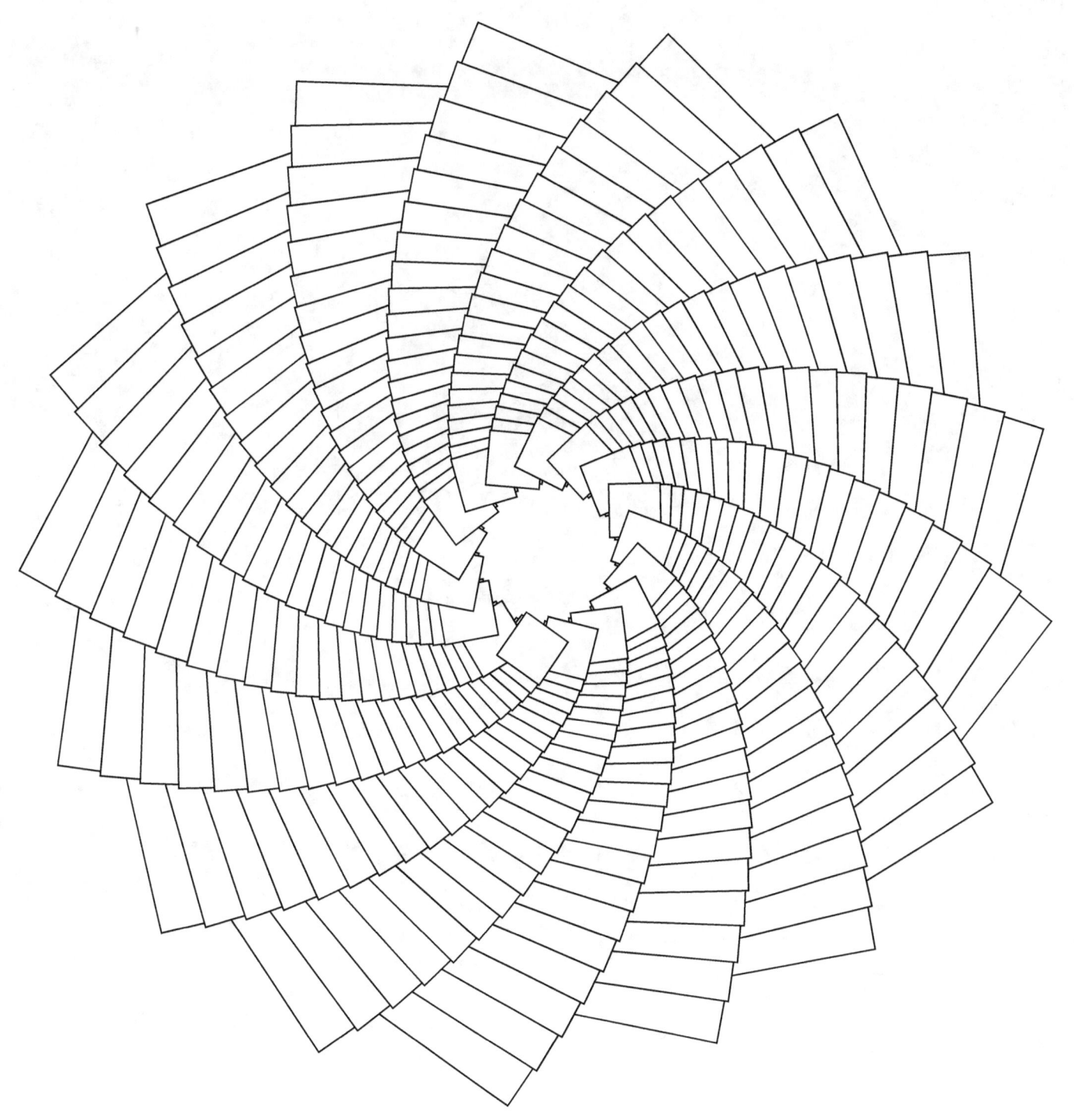

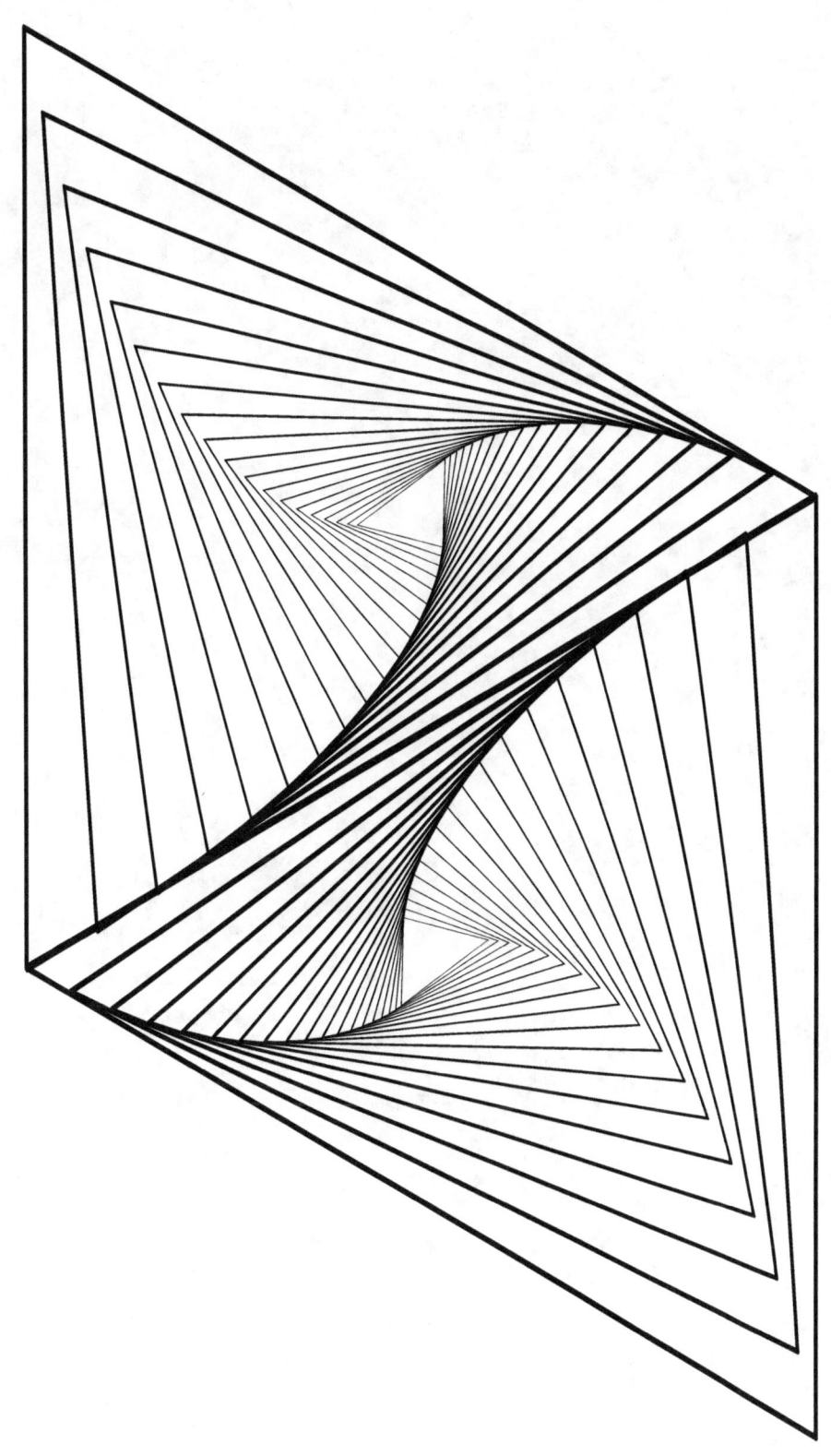

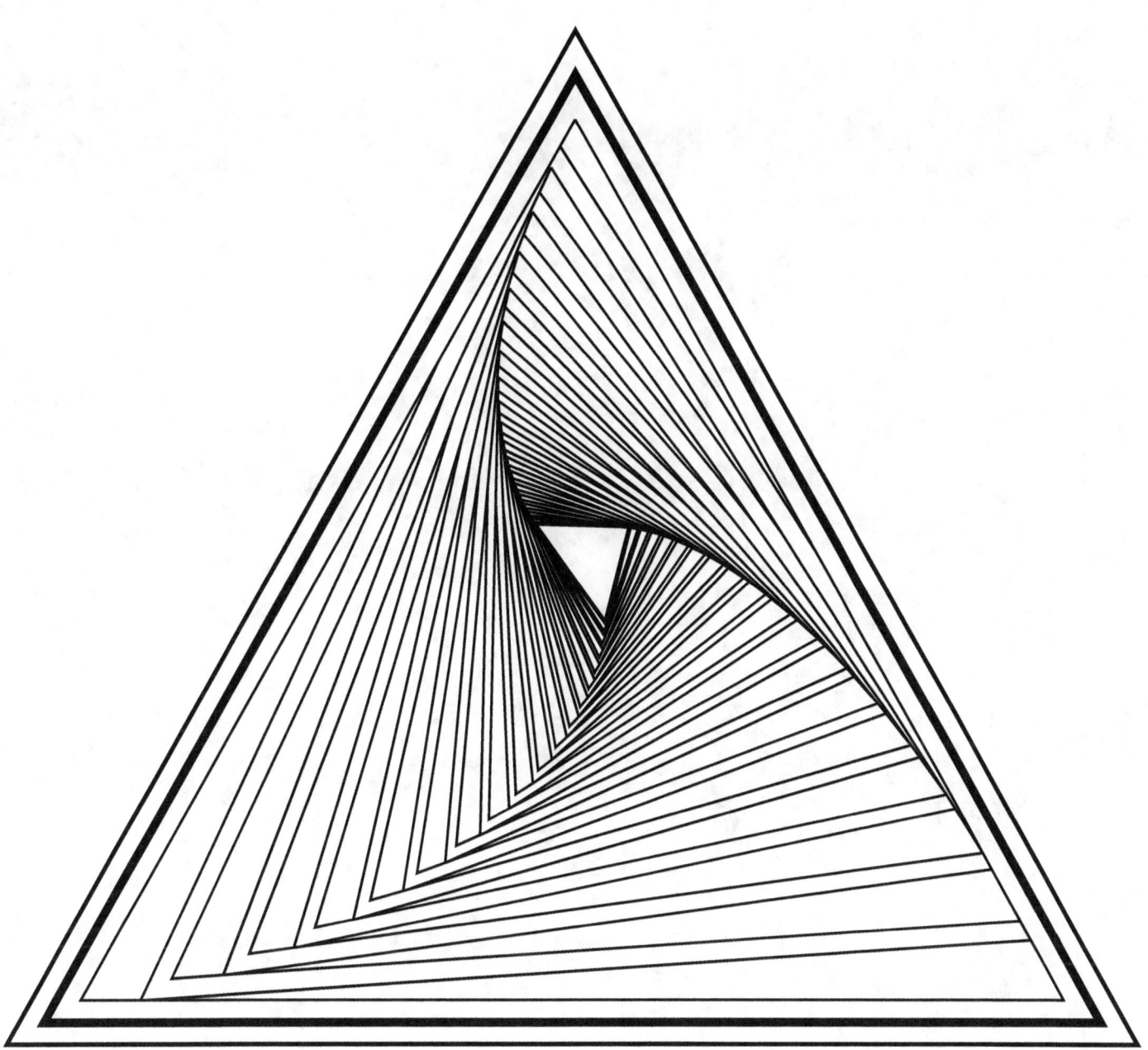

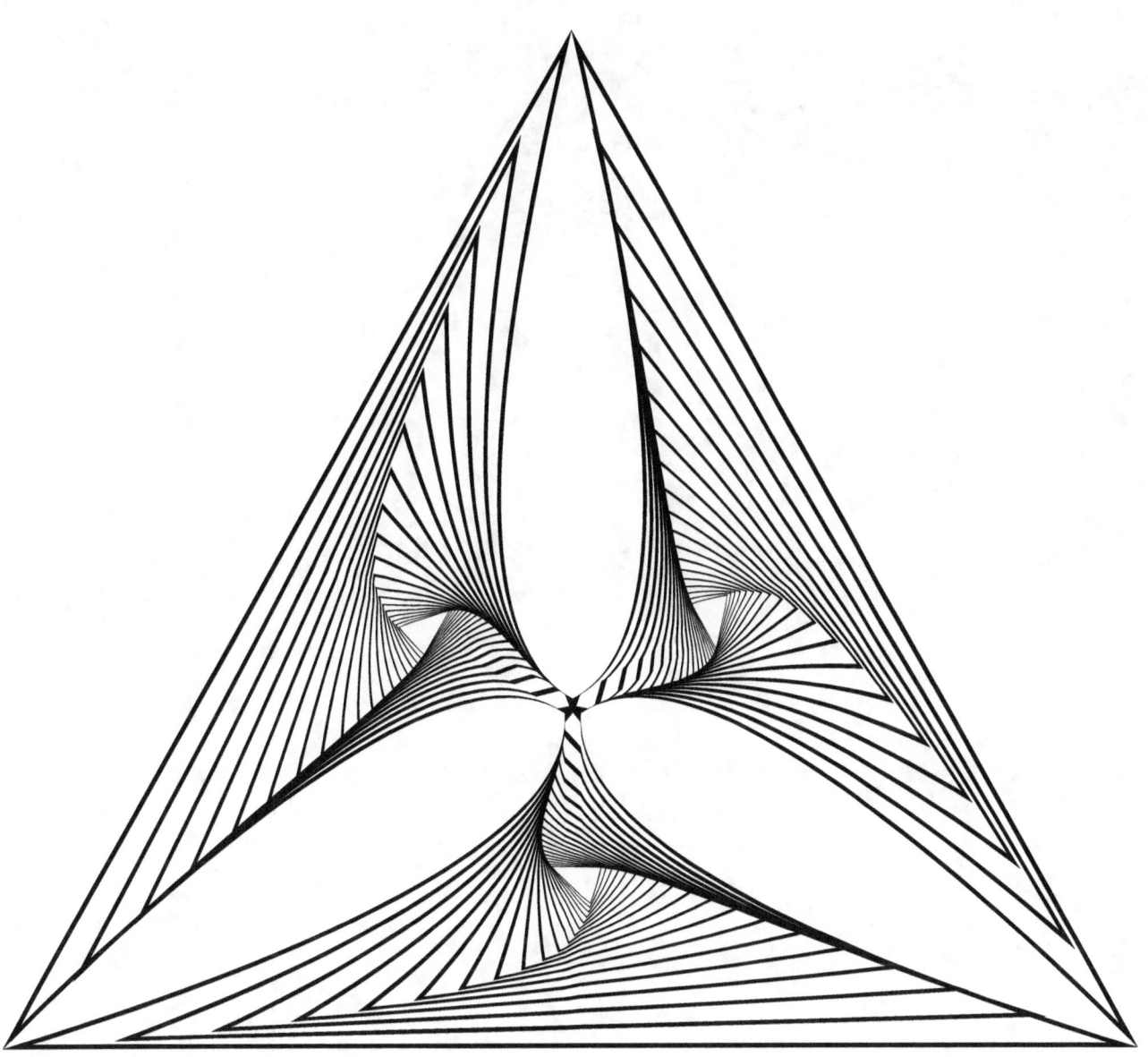

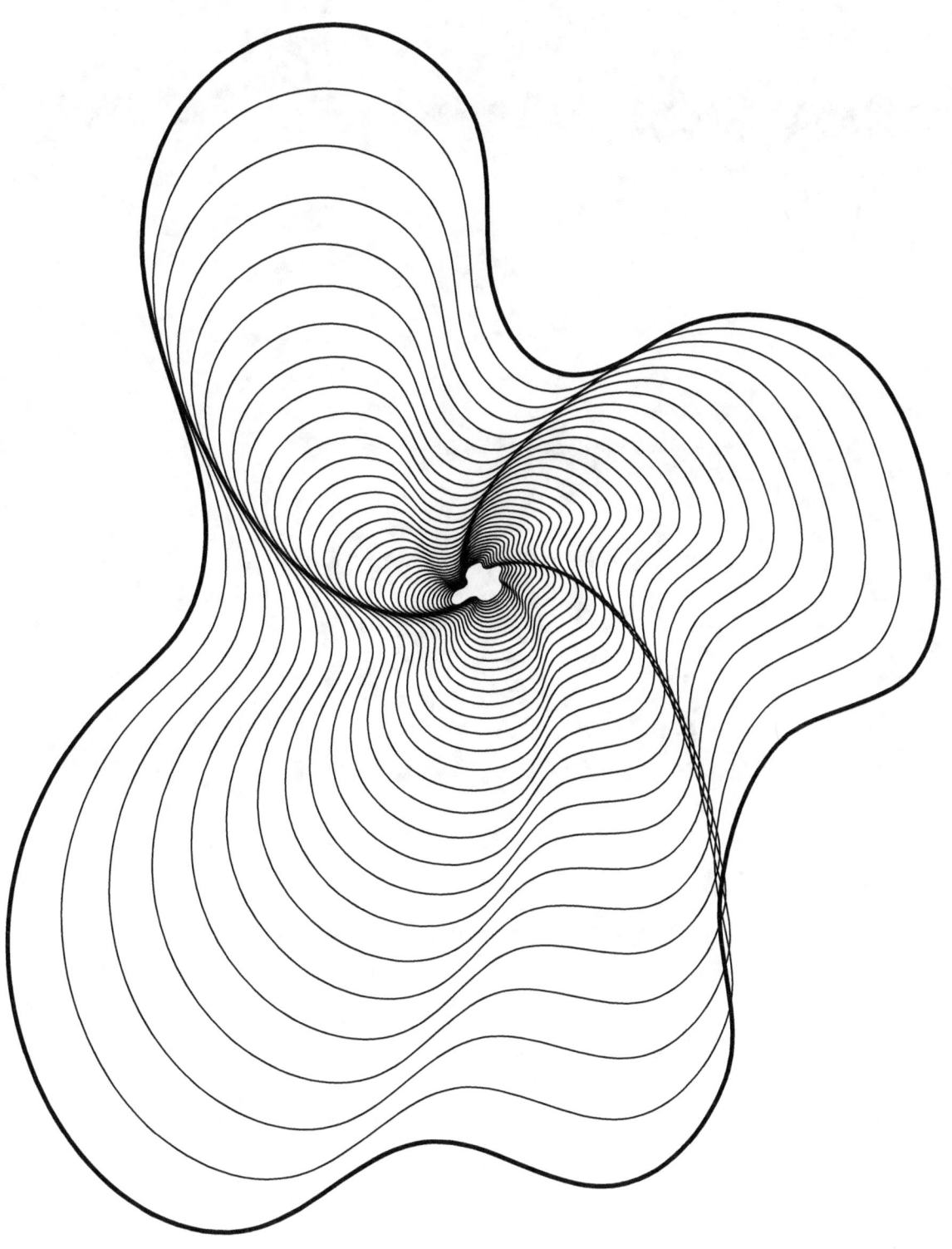

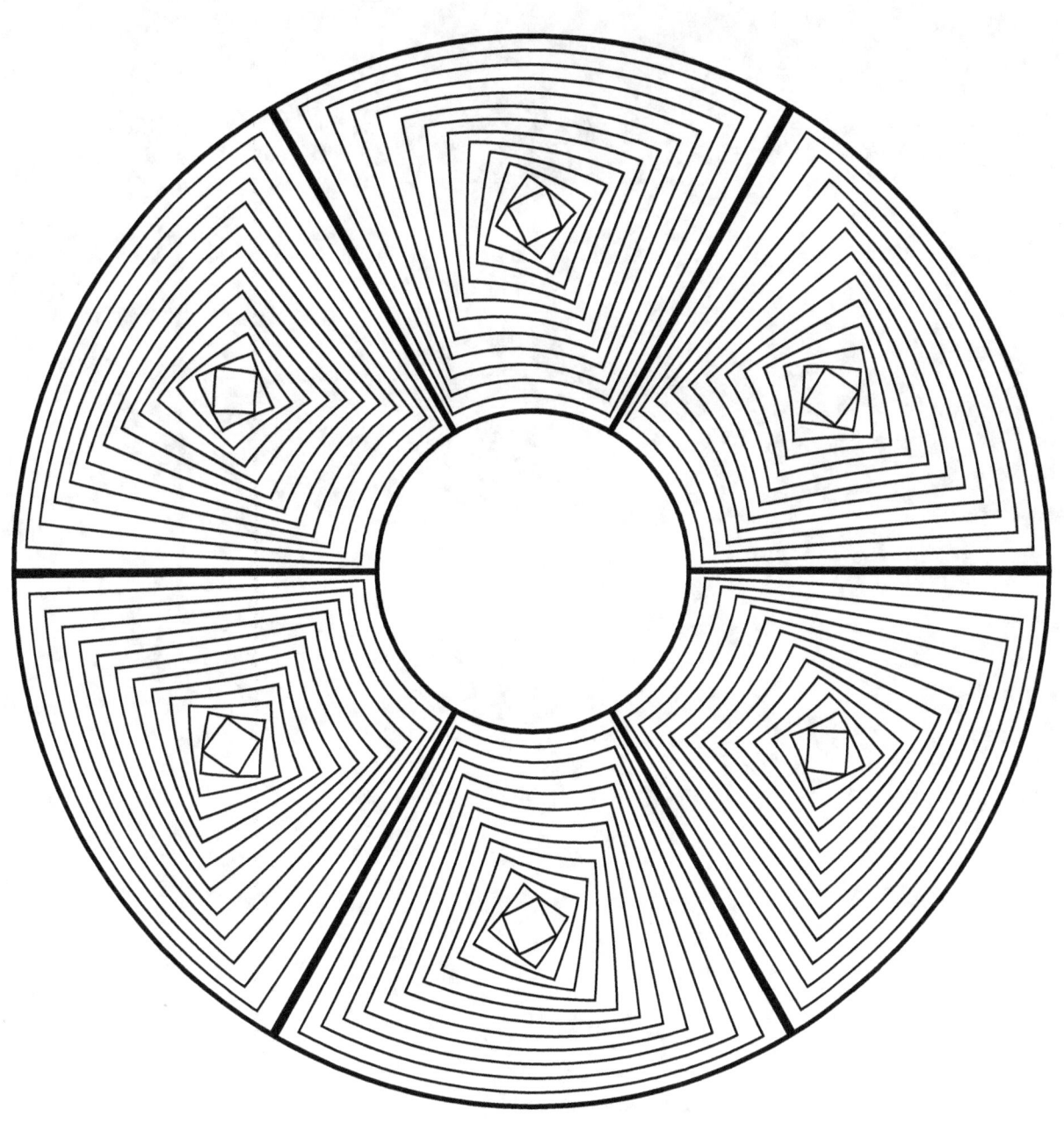

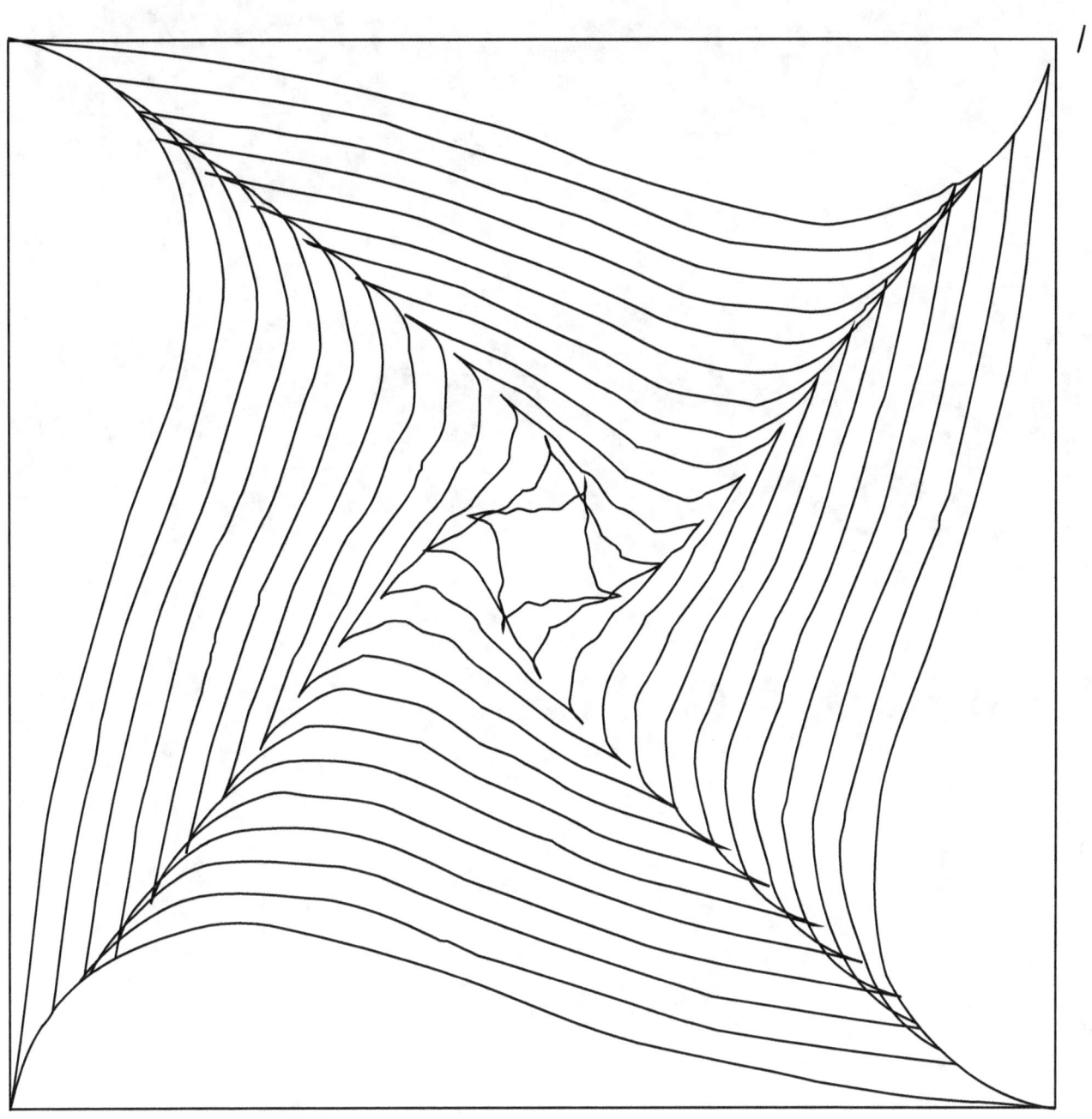

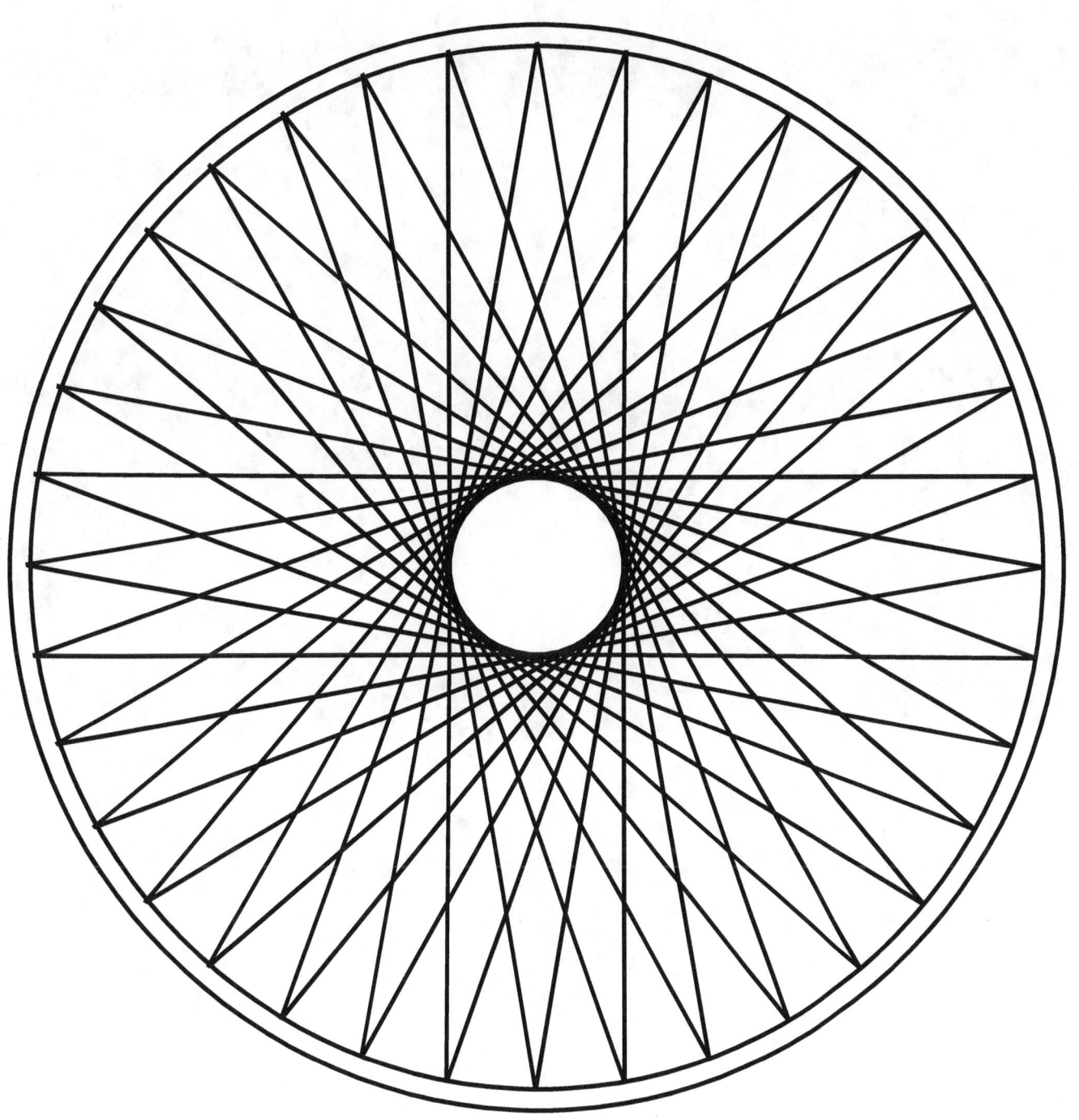

OTHER BOOKS AVAILABLE!

Please go to:
piggybackpress.com

FOR MORE

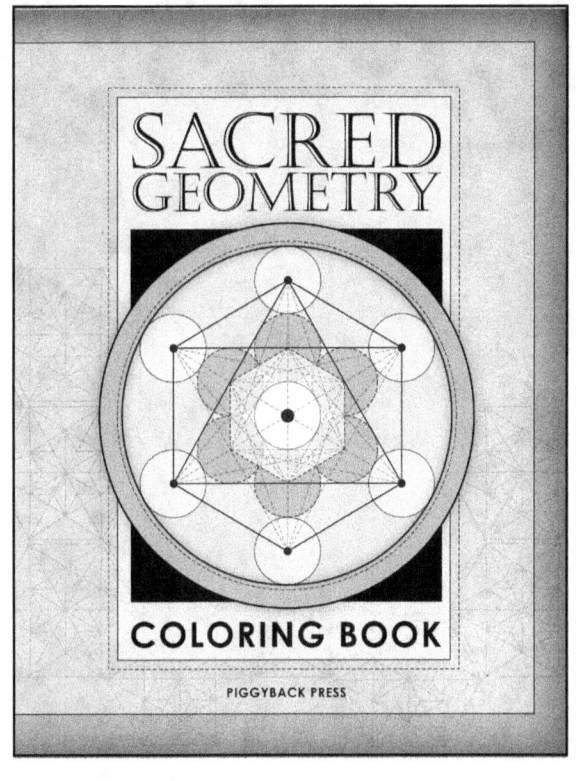

Sacred Geometry Coloring Book

This coloring book was meant for you to enjoy hours of relaxation. Over 60 designs to choose from! Classic symbols include, The Seed of Life, Sri Yantra, Metatron's Cube, Icosahedron, Star Teta-hedron, Torus, the Flower of Life and Vesica Piscis Eye.Explore, have fun and fall in love with coloring these classic geometric shapes and patterns. Sixty-two coloring sheets in all and each design is on it's one individual page.

www.ingramcontent.com/pod-product-compliance
Lightning Source LLC
Chambersburg PA
CBHW081534220526
45467CB00010B/3188